FARMER
IN THE
WIND

CHRIS HARVEY

To order additional copies of this book, contact:
Xlibris Corporation
1-888-795-4274
www.Xlibris.com
Orders@Xlibris.com

Well, Uncle Bill's nephew Ryan wasn't feeling too well, and he was between crops, so he decided to go north and visit Ryan and his older brother Tommy.
He didn't see them that much because he lived down south and was always very busy working on his farm.
Uncle Bill was proud of who he had become and wanted to show his nephews what he did for work and how important a job it really was.

Here is his story........

For all that read this;
May name is Bill, I am a farmer, (so don't be concerned with **spelling or punctuation**).
I'm not your ordinary farmer though, as my pictures will show you.........

In the year 2008 the economy in North Carolina, as well the entire United States, had gone to the dogs.

(*In 2008 the US economy was in a* **recession** *. Experiencing wide spread financial downturns, with a Wall Street and banking crises that had never been seen before in the United States.*)

I was in the swimming pool business and had built a very nice pool for a nice man named Don Reed.

Mr. Reed was very impressed at what a fine job I had done bulilding his pool.
He had told me that it was the finest pool he had ever seen.

As the economy got worse and worse, I had to face the fact that my business, as the old expression goes **"went down the literal drain"**. Leaving me with no job and running out of well needed money.

Fortunately, I approached Mr. Reed about working for him and luckily, he offered me a job working on one of his farms. Now, not being a "real" farmer I said to him, "I've never worked on a farm before", and he replied, "I know, but you sure can operate my kind of **machinery**".

The Wind industry estimates that, as of the end of 2008, it employed 85,000 workers, up from 50,000 a year before—creating a net 35,000 jobs in 2008 at a time when the U.S. lost about 2 million jobs.

Uncle Bill was confused and asked "where is this farm?", Mr. Reed replied, "It's in the beautiful state of **Wyoming**".
Uncle Bill laughed, and asked "when can I start?", "Now" Mr. Reed said!
(*Uncle Bill must have done a good job of installing his pool*).

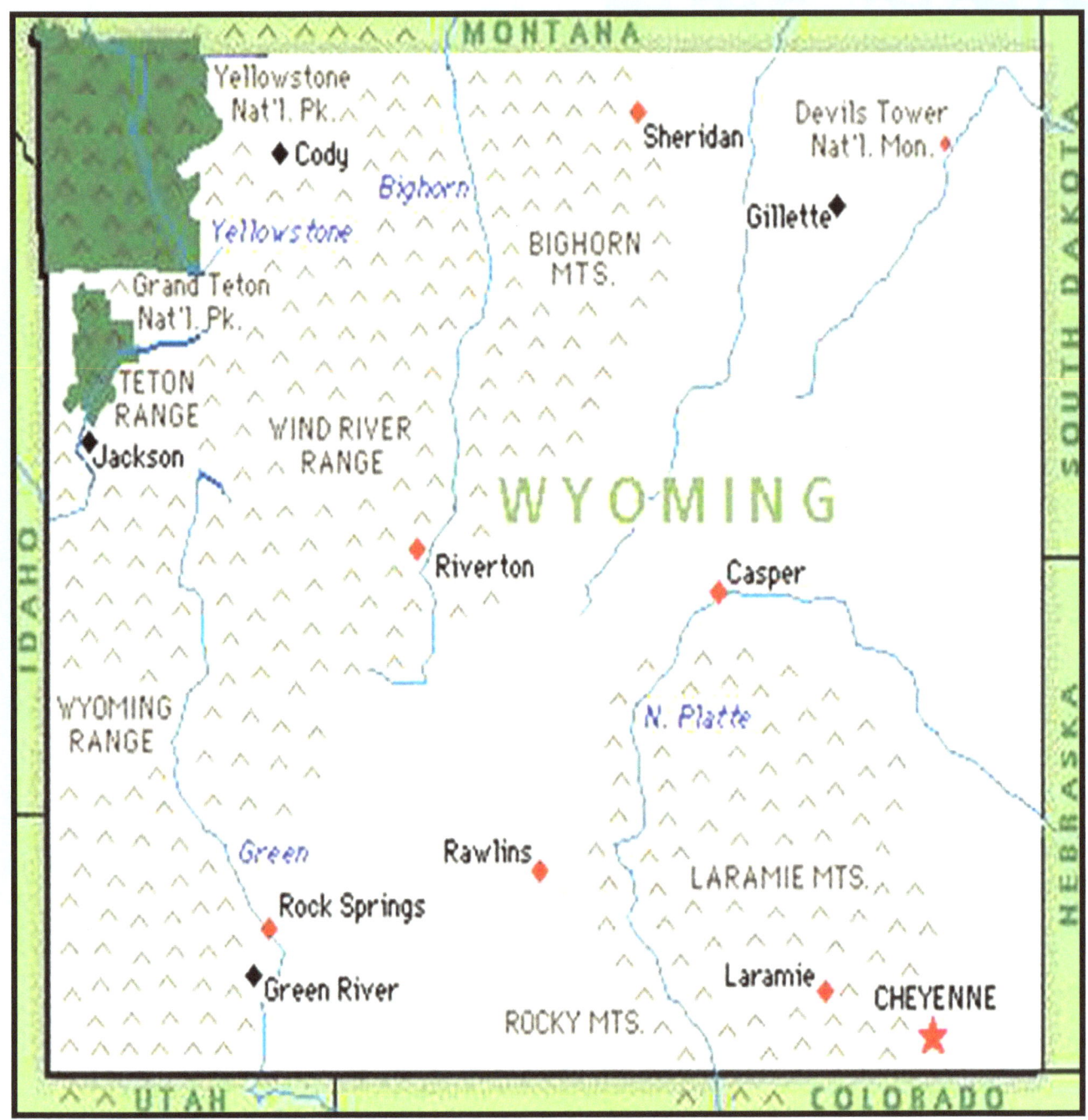

Wyoming produces a substantial amount of wind-generated electricity and the Southern Wyoming Corridor is one of the most favorable locations for wind power development in the Nation. Some of the other states that have major wind farm projects are Texas, California, Minnesota, Iowa and Washington.

So that is how Uncle Bill's career as a farmer got started.....

"What do you do on the farm" Ryan asked?

The first thing I do is dig a huge **hole** in the ground to start the plantings.

These holes must be 30 to 50 feet deep and are filled with cement to anchor the tower.

Next, my fellow farmers plant the rather large field vegetation.

Each blade on the Wind turbines are 110 feet long.

Then we all get together and make them grow to produce their fruit.

" But Uncle Bill, what do you farm?" asked his nephew Tommy.

"Well, I will tell you, we farm the WIND , and it's fruit is **ELECTRICITY**!!"

It takes approximately 3 million dollars and 200 men to build 1 Wind Turbine Tower.

Our farms usually yield about 100-2,500 towers and produce enough electricity to power 3 city blocks per tower in **New York City** for 24 hours a day, 360 days a year.....forever.

"How does electricity come from a wind turbine?" Tommy asked

Wind turbines operate on a simple principle. The energy in the wind turns two or three propeller-like blades around a rotor. The rotor is connected to the main shaft, which spins a **generator** to create electricity.

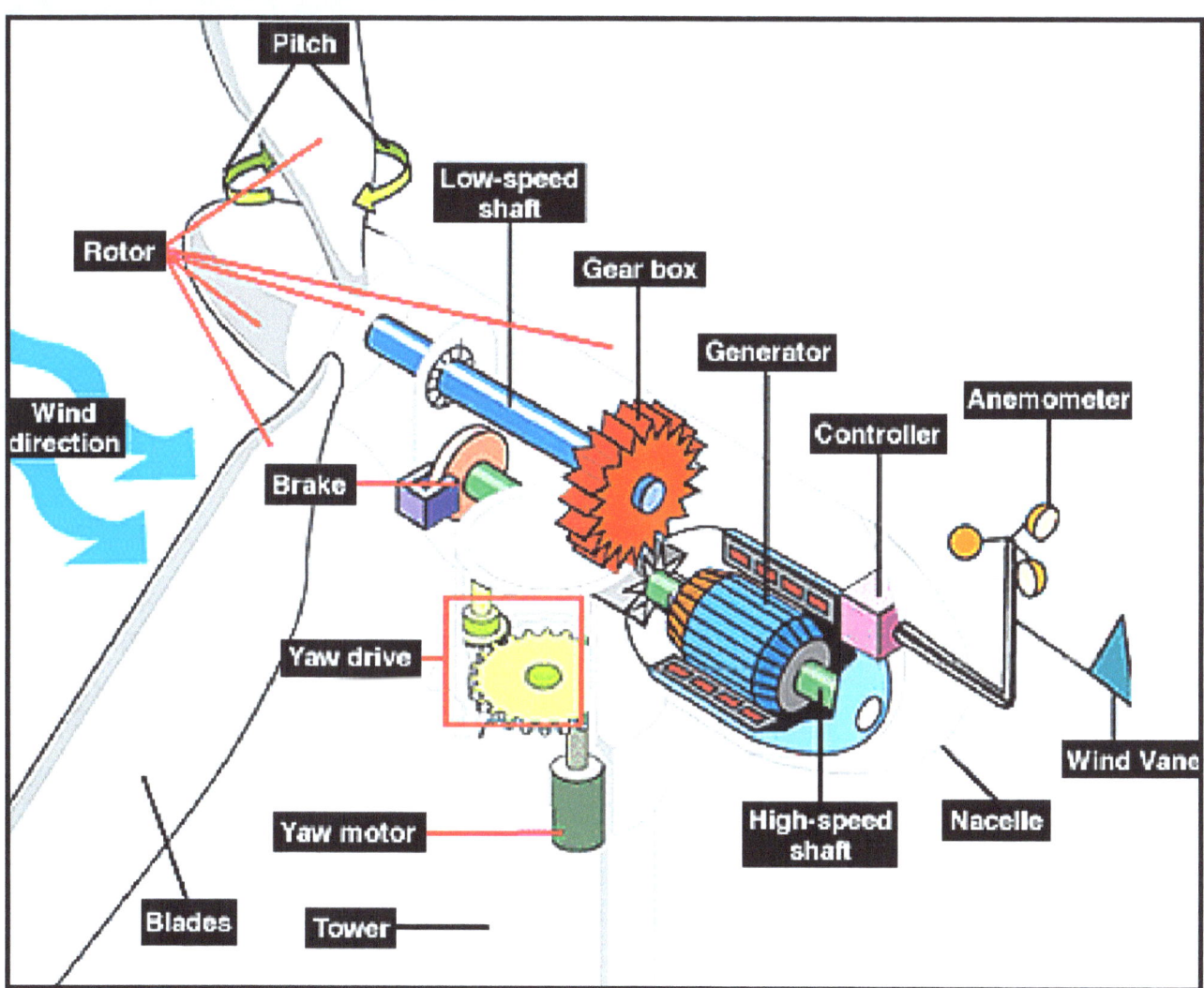

Wind turbines are available in a variety of sizes, and therefore power ratings. The largest machine has blades that span more than the length of a football field, stands 20 building stories high, and produces enough electricity to power 1,400 homes.

We then have to dig long trenches and run the cables that bring the electricity from the Wind Turbines to the **electrical grids** that go to homes and cities.

These special **Trench Excavators** are some of the largest ever built and are able to dig trenches that go on for miles.

Uncle Bill is one of many farmers to work on a Wind farm. Each worker are experts at their jobs.

This is Lee Marshell. He is the nicest guy on the farm. He gave his hard hat to Tommy. That Hard hat has been all over the world with him. He is the best operator of these huge trenching machines in the world!!

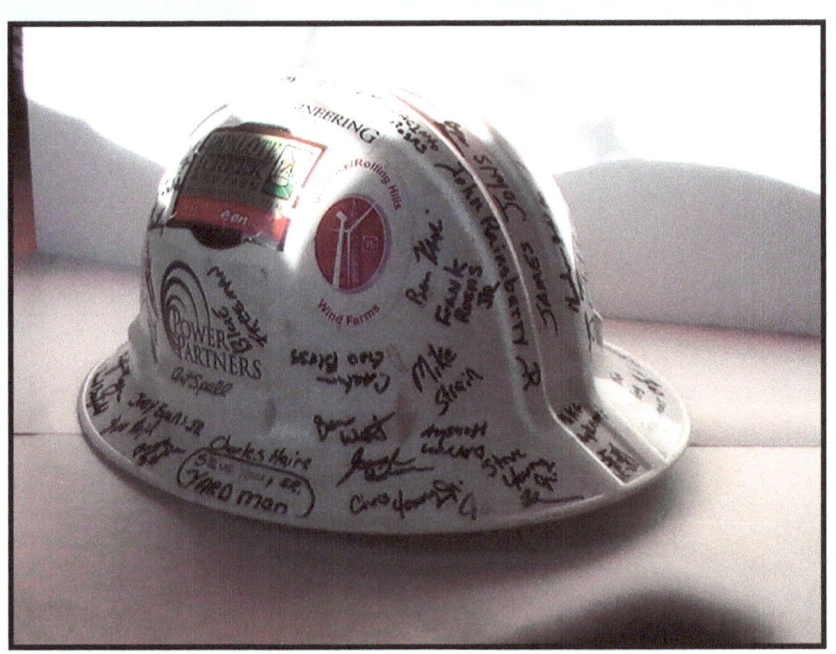

They need hard hats and protective gear as running heavy machinery and building giant towers is very dangerous business.

These Power transfer stations are needed to move the electricity over the electric companies grid network that sends the power to the cities and towns.

"The men in the pictures are a tough bunch, but probably the most honest, hardworking and caring bunch I've ever had the privilege of working with."

"Pay attention in the future, and maybe you will one day turn your lights on and think, hey, this light is powered by a **crazy Farmer in the wind**"

Royalties from the book will go to:
The Ryan Harvey Benefit
PO Box 129
West Harwich, MA 02671

Thank you

To Uncle Bill; The creator of his own life story that needed to be told.
Don Reed; For enabling Uncle Bill to showcase his talents.
Power Partners; For listening to Don Reed.
Maztec Corporation; For acquiring Power Partners and employing the hardworking farmers.

The hardworking farmers; Robert the safety manager (whose helmet that he gave to Ryan saved his life one day), Uncle Bill's crew, Eagle and his crew from Kansas, Jermain from Georgia, the farmers from North Carolina, James from Florida, Doug who gave Ryan a backpack and Tommy a wind turbine model, the only women on the farm (you know who you are) who guided the farmers to sign the hard hats, Big Bobby Morgan from Hendersonville, NC (he's 6'7" 280lbs. Uncle Bill is glad he's on his side), Fran Salazar from Kansas who traded his camper for Uncle Bill's Corvette, Hoss traded his pick up truck for Uncle Bill's older camper, Darren the red neck farmer from Tennessee, Lee Marshell who donated his hard hat to Tommy (that hard hat has been all over the world with him), the other red neck farmers Paul Lewis who is 76 years old and the farmer that drives the biggest trucks on the site Bruce, Steve Young "Yard Man" he gave Ryan and Tommy some cool knives and Pokemon stuff, and all the hardworking farmers that signed the hard hats for Ryan and Tommy.

Glossary of Terms

spelling and punctuation (page 5); It's hard work on the farm...and Uncle Bill wants everyone to work as hard in the classroom as he does on the farm.

Recession (page 6); A general business slump. Making people lose their jobs and adding to companies going out of business.

"went down the literal drain"(page 6); An old expression that means money and expenses being spent are far greater than money being earned.

Machinery (page 7); Uncle Bill can drive and control almost All the large trucks, drills, cranes, and any other piece of equipment on the farm. I have heard him say " give me 15 minutes with any of these large machines and I'll be able to run it". That kind of talent makes him a valuable farmer to have on the job.

Wyoming (page 8); A State in the Northwest United States 470,816; 97,914 sq miles (2503,595 sq km). The capitol is Cheyenne.

holes (page 9); these holes are dug directly in to the hard dirt and sometimes need some type of lubricant so the drill can continue to dig without getting stuck. As Uncle Bill says, "Must apply the lotion!"

New York City (page 14); Is the most densely populated major city in the United States. The New York metropolitan area's population is also the nations largest, estimated at 18.8 million people over 6,720 sq miles (17,400 sq km).

ELECTRICITY(page 13); The time rate of flow of electrical charge, in the direction that a positive moving charge would take and having magnitude equal to the quantity of charge per unit time: measured in amperes.

Generator (pg 15); A machine that converts mechanical energy into electrical energy.

Electrical grids (page 17); Is an interconnected network for delivering electricity from suppliers to consumers.

Trench Excavators (page 17); A digging machine, usually on crawler tracks, and having either a movable wheel or a continuous chain on which buckets are mounted.

"crazy farmer in the wind" (page 21); Actually, Uncle Bill really is not crazy, he just does the hardest, highest and most technical jobs on the farm that most people would never consider doing.

www.ingramcontent.com/pod-product-compliance
Lightning Source LLC
Chambersburg PA
CBHW050429180526
45159CB00005B/2474